ECHOLINK FOR BEGINNERS

BRIAN SCHELL

CONTENTS

FOREWORD

When I wrote the original "Echolink for Beginners (2014)" a few years ago, it was my first amateur radio-related book. I saw a need to clarify what Echolink was and wasn't, and for a plain English explanation of how to get it set up for the first time user. Although it was criticized for not going far enough, a criticism in which in hindsight I agree, it quickly became one of my best-selling books. A **lot** of people were interested in learning about Echolink.

In this new edition of the book, I wanted to answer some of the critics by "going further," and expanding on the original "for beginners" concept and put more detail on what I consider more advanced topics, such as radio interfacing.

There is an entirely new section added to the book called "Sysop Mode," which details interfacing the very common Baofeng UV-5R radio to a computer with a SignaLink device and setting up the software to create an Echolink "Linked System." Baofeng radios and SignaLinks are commonly available and as inexpensive as radio equipment gets, and the procedures involved can be easily adapted to a wide variety of other equipment.

In addition, I have looked at software enhancements to Echolink, explained the use of DTMF, and spent more time explaining the fine-tuning of some of the Echolink software settings.

Although the basic Echolink software hasn't changed even a little since the previous edition, I think you'll find that the Echolink community is as active and as vibrant as ever, keeping the conferences and repeaters active and fun, 24x7.

Brian Schell, KD8OTD

WHAT IS ECHOLINK?

E cholink is a computer application or mobile app that allows a licensed amateur radio operator to talk to the world through various modes connected to the Internet.

The beauty of Echolink is that it allows these communications in several ways:

1. You can use your handheld radio to connect with a local repeater that is Echolink enabled, then use that link to talk to people anywhere in the world.
2. You can connect your own local "base" radio to your PC and then use a handheld radio and a simplex connection to talk through the Internet to other hams worldwide.

Both of these methods are similar to using a regular FM repeater, in that you use a regular radio to connect and talk to other people remotely. But that's not all:

1. Another very popular use of Echolink is to just plug a microphone into your PC and talk to hams

around the world through the Internet, very similar to using Skype or another VOIP (Voice Over Internet Protocol) system.

2. You can even listen and talk to hams worldwide through your wifi-enabled smartphone or tablet with no additional equipment at all.

It's just crazy flexible, and all these options are what tends to overwhelm people. It's really very easy to set up Echolink the first time, and then it's just a matter of how far you want to go with it.

The first major requirement is that in order to use the Echolink system, you *must* be a licensed amateur radio operator ("ham"). If you aren't a ham radio operator, then that's the necessary first step. In the USA, there are three main "levels" of license: Technician, General, and Amateur Extra. Technician is the beginning level, and it's not that hard to attain. The Technician-level license is all you need to do everything Echolink offers. The higher-level licenses allow you to do other fun things with radio, but they won't add anything as far as Echolink is concerned.

Is Echolink Really "Radio?"

There has always been controversy over Echolink by purists who insist that it "isn't real radio." There's really no reason to debate this, it's really just a matter of how you use it. If you are holding a handheld HT radio in your hand and using it to connect to local repeater that connects through the Internet to a repeater in Sweden that is being used by a Swedish ham who is also holding a radio in his hand, it's hard to argue that this isn't "real radio." It may be Internet-assisted, but there are still radios and over-the-air transmissions involved.

On the other hand, if you are sitting at your desk, talking into a microphone connected to your computer, with no radio in sight, and you are connected through Echolink to a user in Toronto who is running the Echolink app on his Android phone, is that "real radio?" Probably not; there's no radio transmission anywhere involved. It's really not any different than talking on Skype.

The reason Echolink users need to be licensed hams is that you *don't know* what the person on the other end of the connection is using. If you are on the computer and connecting to a ham in Sydney Australia, you don't know if he or she is on a phone, tablet, computer, or sitting in front of a radio system that cost more than your car. You just can't tell.

But I'm Not a Ham!

A great place to start if you aren't licensed is the ARRL Ham Radio License Manual. It's the book that gets most hams started. With that out of the way, everything else in this book will assume you are a licensed amateur.

Similar, but Unrelated Modes

Echolink isn't the only Internet-assisted way of communicating long distances with radios. You may have heard of the following two modes:

D-Star – D-Star is another very popular method of mixing amateur radio activity with Internet connectivity.

Whereas Echolink is primarily driven through software installed on a computer or mobile device, D-Star relies on hardware decoding through the AMBE chipset installed in certain radios or equipment. This requirement for specific physical hardware makes D-Star a more expensive alternative than Echolink, but it simplifies usage somewhat since there is no need for a computer. The (digital) radio connects through a special D-Star repeater directly, eliminating the need for the computer connection. Check out the author's companion book, "D-Star for Beginners" for a comprehensive explanation of the system.

IRLP – Is short for the *Internet Radio Linking Project*. See more at http://www.irlp.net/ It's very much like Echolink in that it uses standalone computers to connect radios through the Internet. It generally requires a special interface board, and therefore is a little more hardware-intensive than Echolink.

If you have heard of or have used either of these two methods of long-distance Internet-supported radio communications, you will find Echolink to be along the same lines. If you find that you enjoy Echolink, these modes might be a fun way to expand your hobby in the future.

PC INSTALLATION

The first step is to install the software.

On a Mac:

There is no officially supported version of Echolink for OSX. At one time, there was an unofficial way to use Echolink on a Mac, called EchoMac, found at http://echomac.source-forge.net, but the most recent version of the software has a 2006 date and appears to be abandoned. I am running OSX 10.11 El Capitan, and although the software will install and run, it won't connect to anything. Your mileage may vary, but I tend to tell people that Echolink just won't currently work on a modern Mac.

Another option is to use Bootcamp, Vmware, Parallels, or some other virtualization software to run a version of Windows on your Mac, and then install the Windows version of Echolink. It's not a good solution, but right now, it's the only way to get a current version of Echolink to work on OSX/Mac.

On a Windows Computer:

1. Make sure your computer has the necessary hardware. You need an Internet connection. It doesn't have to be anything fast or outstanding, even old-fashioned dial-up is usually fine for Echolink.
2. You'll need a microphone. There are a thousand varieties of microphone, but for Echolink you don't need anything super-high quality. Most laptop computers have a microphone built-in, and these are often fine for Echolink. If you have a microphone of some kind already, then proceed with the rest of the installation. When we get to the "Testing" chapter, you'll find out whether you are happy with the quality or not. If you don't have any microphone, then I recommend the Logitech ClearChat Comfort / USB Headset H390 which includes a headset and microphone. It works great for me.

That's all you need equipment-wise for a very basic setup.

The next step is to download the software. Go to http://echolink.org and click the "Download" link. You'll see there that you can't just download the software directly, you need to provide your call sign and email address. Fill in the two blanks and you'll be sent to a download page for your PC. Click on the link for whatever the current version of

Echolink is (version 2.0.908 as of this writing), and it will download to your computer. When the download is finished, run the file and install it with all the default options.

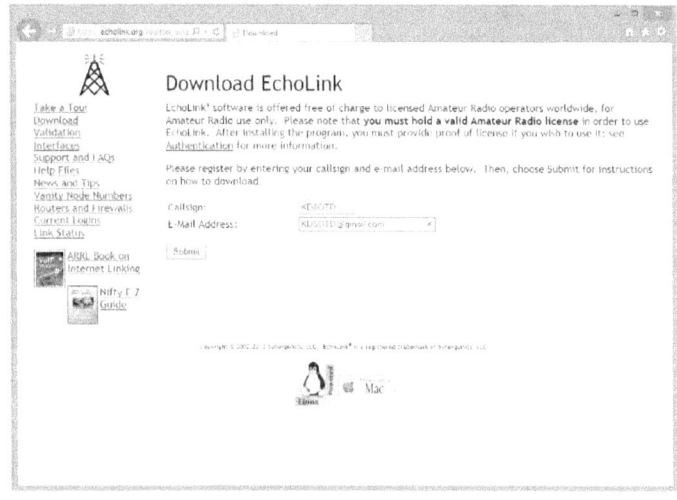

Figure 1 Echolink.org Download Page

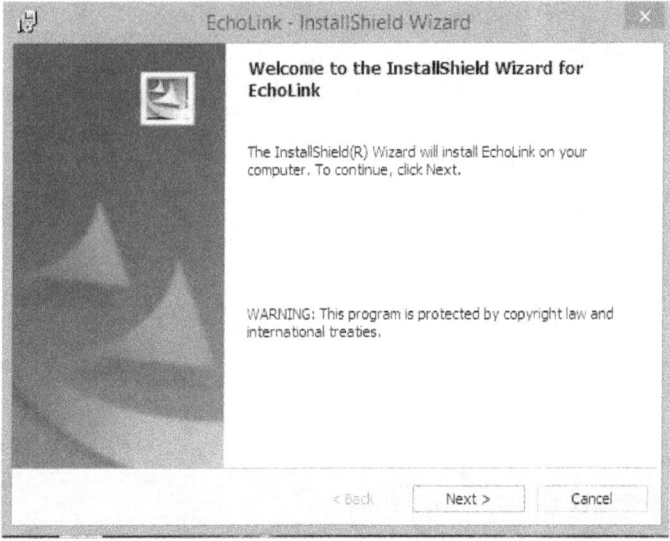

Figure 2 First Installation Step

Click "Next" when you see the dialog window pictured in figure 2.

You may or may not see the following dialog:

Figure 3 Firewall Warning

Assuming you are in a safe environment such as your own home, then accept the default options and click on "Allow Access."

The next screen starts out "Welcome to Echolink," and continues with a for-licensed-hams-only warning. Just click on "Next," since there's nothing to do here anyway:

Figure 4 Welcome to Echolink Setup

Next, you need to make a decision. Are you going to connect a real physical radio to your computer, or are you going to just talk into Echolink through your computer? If you have a radio and interface, then you will want to (eventually) click on "Sysop." For now, especially if this is your first time setting Echolink, I would recommend going with the "Computer User" option. It's just much simpler in the beginning. We'll come back to Sysop mode later in the book. For now, let's assume you are using the "Computer User" option for the sake of simplicity:

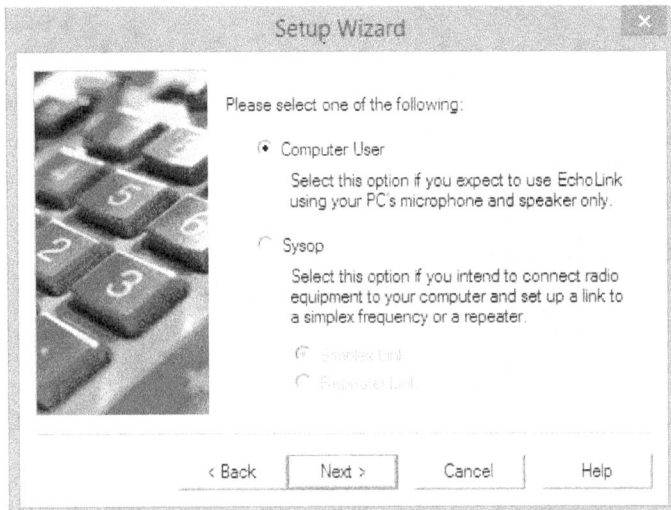

Figure 5 Computer Only or Radio Too?

Next, Echolink needs to know a little about the speed of your Internet connection. Echolink is basically just encoded audio, so it doesn't require a fast Internet connection at all. On the other hand, if you have a fast connection, Echolink can take advantage of the extra bandwidth:

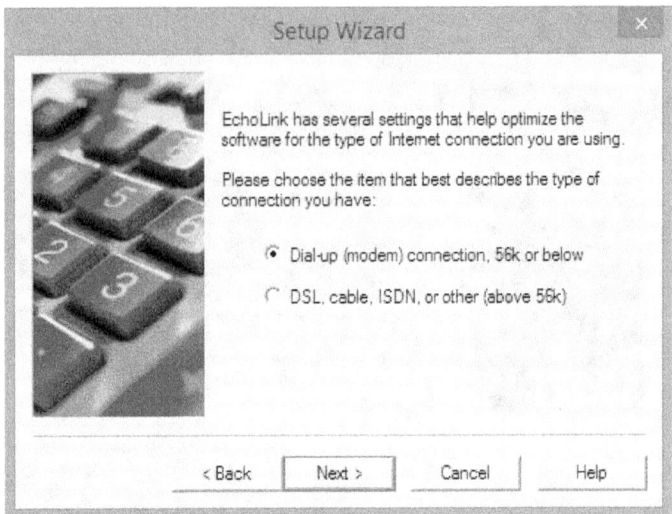

Figure 6 Internet Speed Options

Next, we start getting personal. Enter your amateur radio call sign, first name, and other information. You also need to set up a password for your account. Just make one up, but make a note of it. Your name, call sign, and location can be seen by others using Echolink:

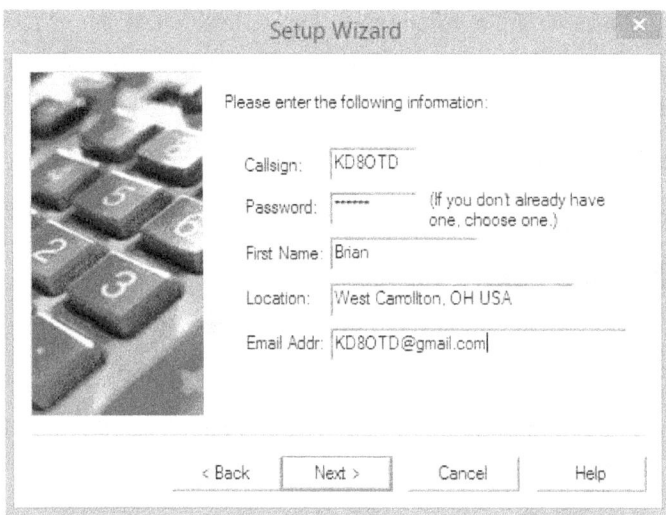

Figure 7 Enter Personal Information

Next, choose your region:

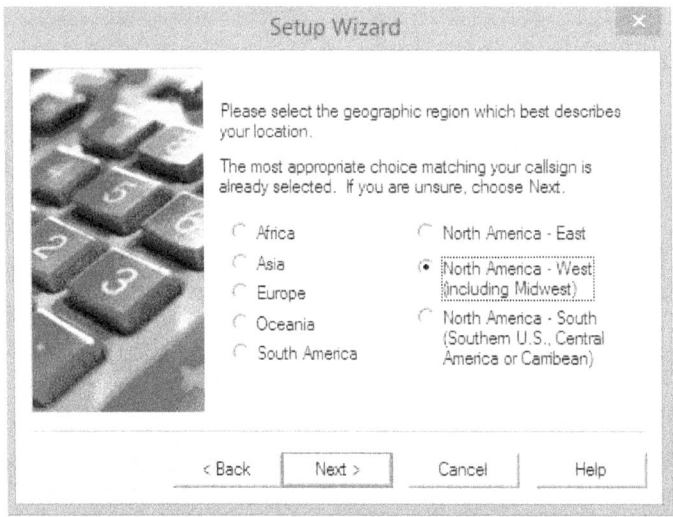

Figure 8 Global Region

Next, we get a dialog titled "Firewall/Router Tester." Most likely, your computer goes through a router to get to the Internet. Click on the "Firewall Test" button and see what happens.

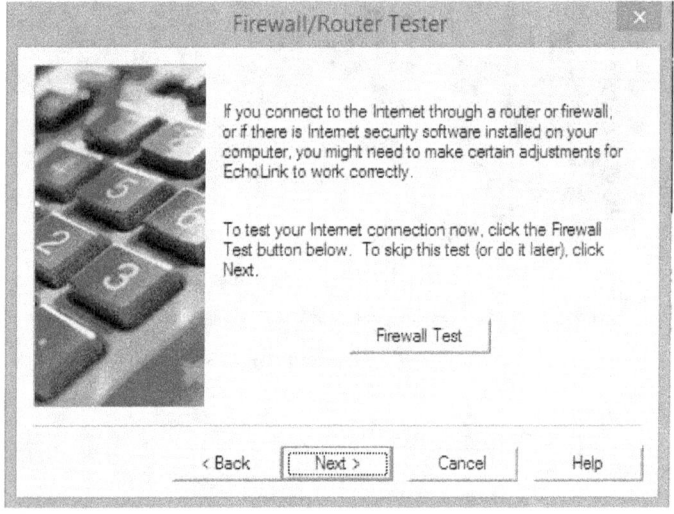

Figure 9 Firewall/Router Test

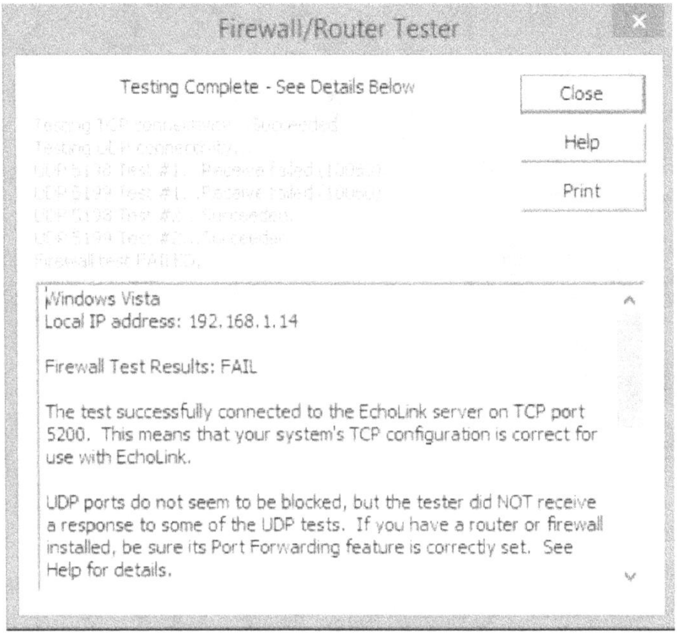

Figure 10 Firewall/Router Test Results

Most likely, your system will pass the test. Echolink uses certain software "ports" to stream information through the Internet, and these ports must be open to Internet access.

As you can see in the results in Figure 10, it reports that my system failed the test. As it turns out, my system works fine anyway. If your system passes the test, then great. If it fails, continue with the steps that follow and see if it works anyway. If your router fails this test, and Echolink actually *doesn't* work, then you will need to open the ports on the router. You will need to check the manual for your router for instructions on that.

At this point, the next screen explains that the installation of the software is finished, but you still need to verify your call sign with the Echolink servers. Follow the instructions that follow on the next screen.

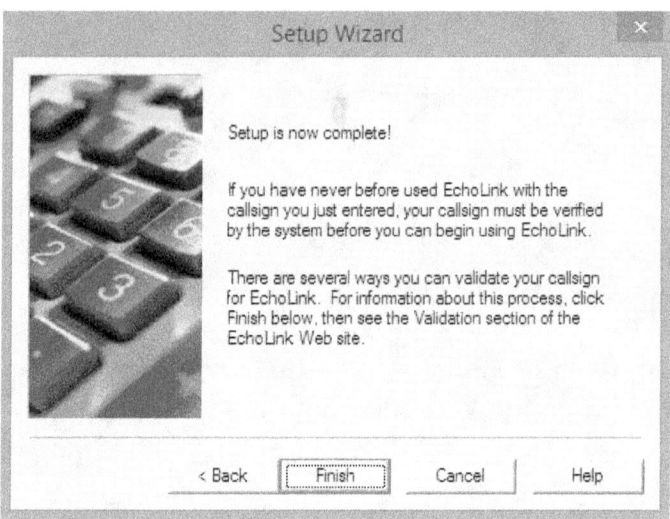

Figure 11 Installation Done

And the installation is done. Once you have the validation process sorted out and completed, you'll be presented with the following screen:

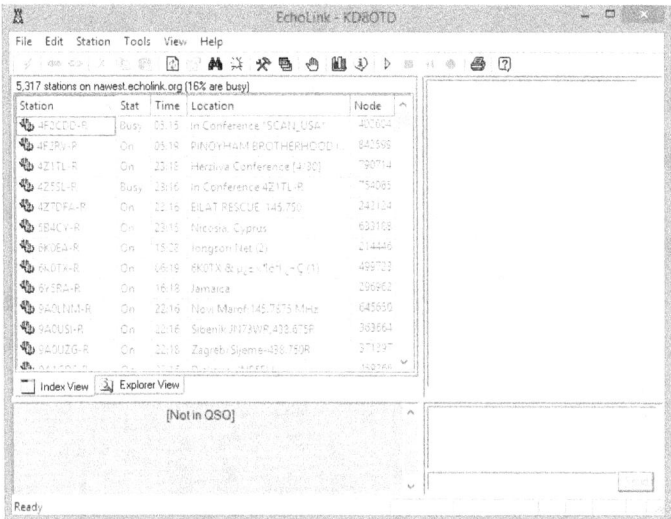

Figure 12 Echolink Main Program Screen

So your Echolink software is installed and seems to run. The next step is to set up various options and get it to work with your microphone.

From the main program screen (Figure 12 again), pull down the "Tools" menu and choose the "Setup" option. You'll get a dialog that looks like this:

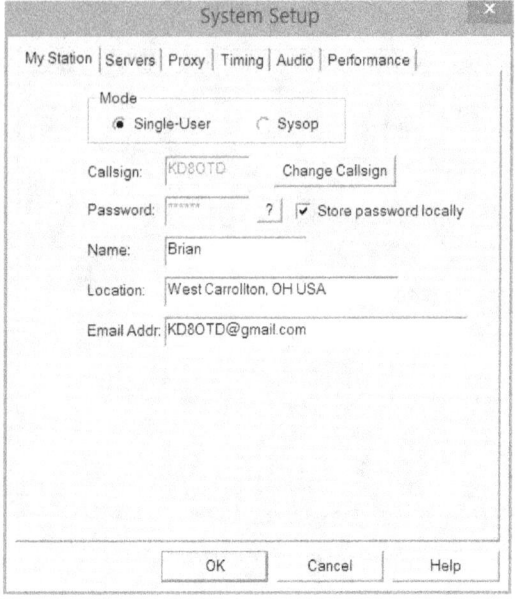

Figure 13 My Station Details

This is all the stuff that you already entered when you installed the program, so you shouldn't need to do anything here. If you see something wrong, change it now. When you get to the more advanced stages later on, this is where you'll switch from Single-User mode to Sysop mode. Notice the tabs along the top: My Station, Servers, Proxy, Timing, Audio, and Performance. Let's look at each of these in turn now.

Servers

The Servers dialog allows you to change which Echolink servers you are connecting to. Remember when you entered

your "Region" in the installation? This is where that information went. Assuming you entered the correct region then, you
shouldn't change the default settings here.

If, on the other hand, you move to a new region, you can
change to the appropriate new servers here.

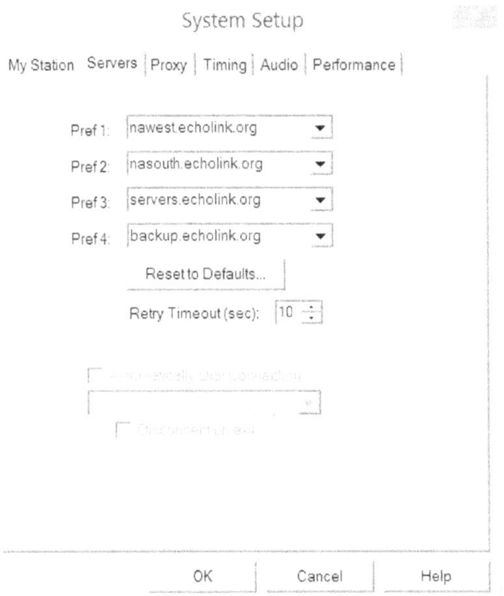

Figure 14 Server Selection

Proxy

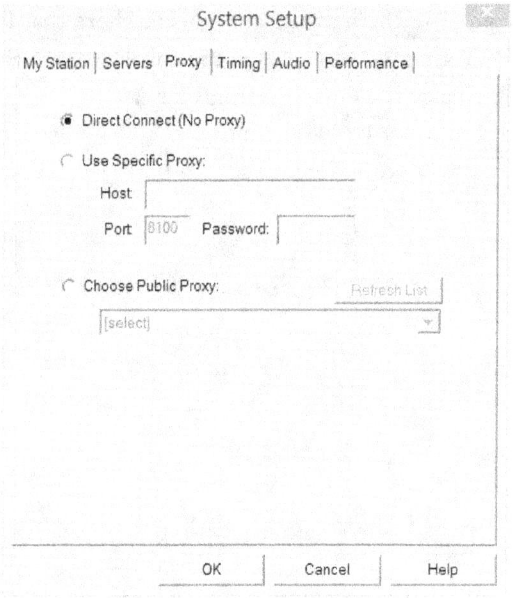

Figure 15 Proxy Server Settings

Depending on how and where your computer is hooked to the Internet, you might need to use a proxy server. If you don't know anything about a proxy server, then you probably don't need one. If you do, then check with your network administrator about what settings you need to enter here.

Timing

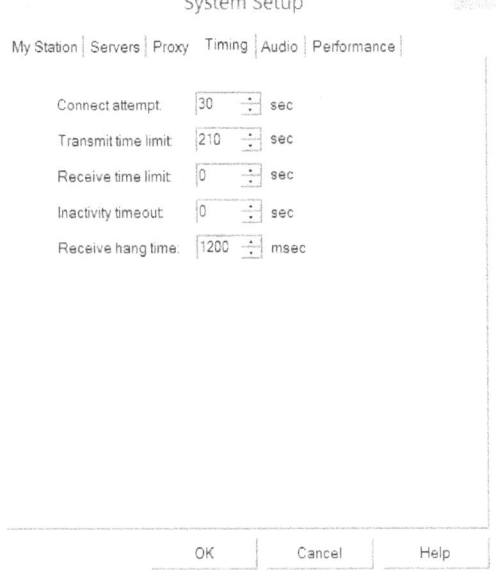

Figure 16 Timing Settings

Sometimes things don't always work as they should, and you don't want your system locking up if something goes wrong with the Internet somewhere down the line. This screen allows you to set timeout limits on various functions. For example, if you cannot connect to a target system within 30 seconds, the system will time out and give you an error message rather than trying indefinitely.

The default settings are usually good, but they can be adjusted if you find you have the need.

Audio

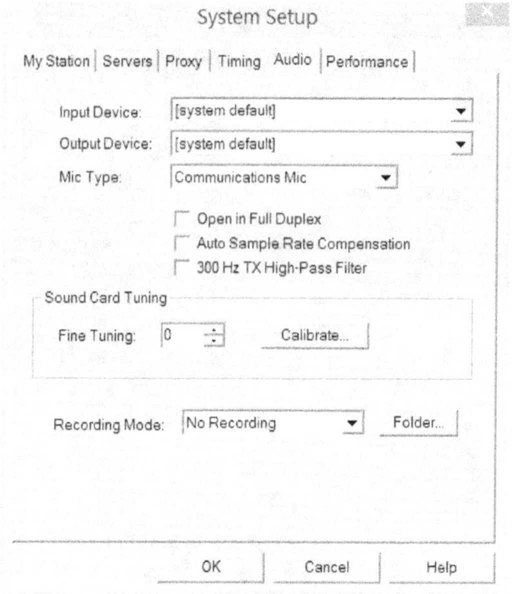

Figure 17 Default Audio Settings

This is the main dialog that you need to make sure is correct. This is where you tell Echolink which microphone and sound output device to use.

If you are running Echolink on a laptop with a built-in microphone, then the defaults here are probably fine, at least while you're testing things. If you are going to use a headset, like the Logitech headset I mentioned earlier, then you need to make sure it's selected in both the "Input Device" and "Output Device" like this:

Figure 18 Logitech USB Headset Selected

When you pull down the selections for the Input and Output devices, it should be fairly obvious which sources you want to select. If your headset or microphone doesn't appear on the list, you should look at the Windows Control Panel and make sure you have a current driver for your device.

When you have the right devices selected, click on the button to "Calibrate" to see how the device works with Echolink. If you have the proper sound drivers installed that match your device, there shouldn't be any issues.

Another option is the "Recording Mode." You can choose to record everything that's said in your conversations. It's generally not necessary, and the recordings do take up hard drive space, so this isn't something most people do, but if saving your conversations is of interest to you, try playing with the settings here.

Later on, when you get started actually listening and talk-ing, if you run into a situation where you get no sound, or people on the other end can't hear you, this is the screen you want to come back to. Even if everything is plugged in correctly, make sure your software is looking in the right place for your microphone and speakers.

Performance

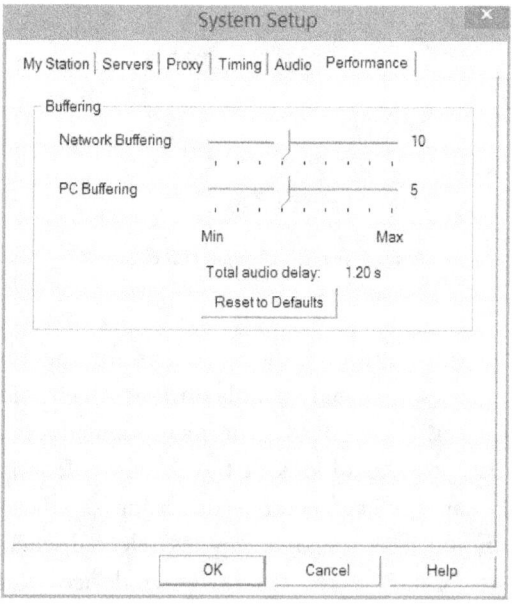

Figure 19 Performance Options

Again, the defaults here are usually fine for most users. If you find that your system or connection is particularly slow, you can try changing the buffering. This means that the audio coming in is "buffered," or held in memory for a few seconds before it is played. Doing this helps reduce "skipping" or lag due to a too-slow Internet connection. Unfortunately, it also makes your response time slower in your conversations, which can be annoying.

MOBILE DEVICES

At the time of this writing, in the middle of 2015, you can get Echolink software for iOS and Android devices. Both the Apple and Android apps look essentially the same, and have the same feature set. With iOS/Apple, there isn't a specific version for the iPad, so you'll need to download the version made for the iPhone.

In this chapter, we will walk through the installation on an iPad, but the iPhone or Android screens will be identical.

If you want to run Echolink on your iPhone, iPad, or Android device, you have it easy. For an Apple tablet or phone, just go to the Apple App Store, or if you have an Android tablet or phone the Google Play Store. Search for "Echolink" and download the app in the usual way. It's free and installs like every other phone or tablet app.

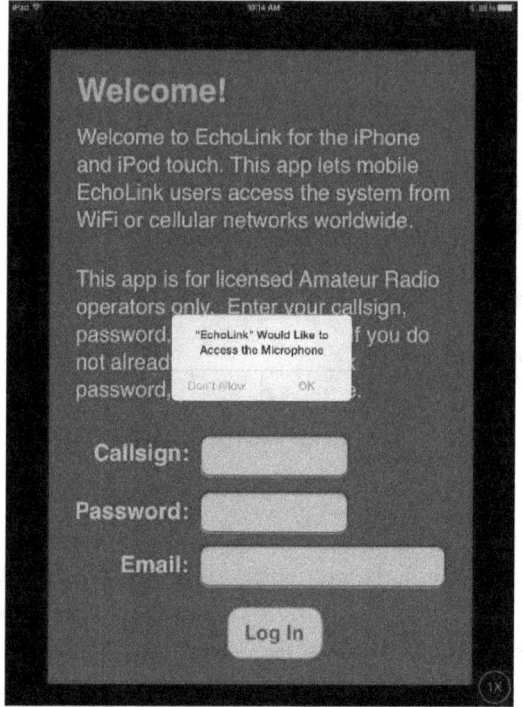

Figure 20 iPad Welcome & Microphone Request

The first screen you'll see looks like the one above. Depending on your security settings, Echolink may give the message "Echolink Would like to Access the Microphone." Since you'll be needing the microphone to talk, you must click "OK" to allow access.

Next, it asks for your call sign, password, and email. Fill in the blanks with whatever you submitted on the computer version. If you didn't do a computer setup, that's fine; just fill in the call sign and email, then make up a new password-- but write it down, you'll need it again.

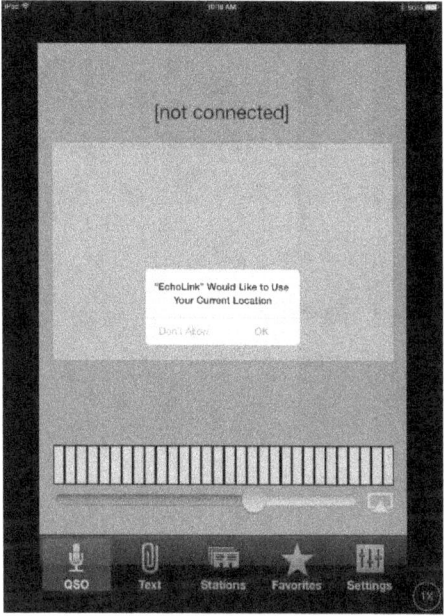

Figure 21 Location Permissions

Once you're through that screen, again depending on your security settings, it may ask for permission to use your current location. All this does is list your city or town next to your call sign on the various lists and menus. It's usual to allow this as well, since this information is easily looked up online through your call sign, but it's not absolutely essential if you have a reason not to allow this.

Figure 22 Station List

Now you are at the main Echolink screen, the Station List. The first option on the list is "**ECHOTEST** Audio Test Server." Click it. You'll see a screen like this:

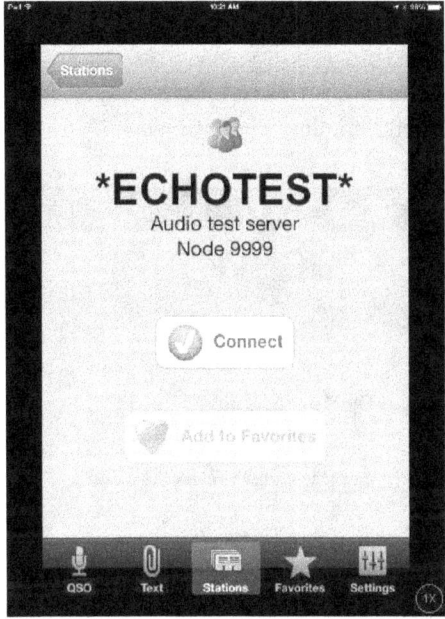

Figure 23 Echotest

The Echotest server is an automatic testing system that lets you record your voice, send it through Echolink to their servers, and then listen to it the way other people will hear you. It's a great way to make sure your microphone is working and that all the volume and level settings are set properly. No human besides you will ever hear this, so feel free to experiment and practice talking at various volume levels. When you talk to real people later, this is what you'll sound like, so if you can't hear yourself, you will need to speak louder. If it hurts your ears, you'll need to tone down a little. Experiment!

Click on the green "Connect" button, and a screen like the following will show up. You'll hear a voice explaining what the Echotest server does. When the voice is done explaining, you'll see a button marked "Transmit." Click it. You'll get a picture of a microphone. Talk into your mobile device and tap on the screen when you are done. Assuming everything is working correctly, you'll hear whatever you just said repeated to you. If it worked, you have everything you'll need to "work the world" with Echolink.

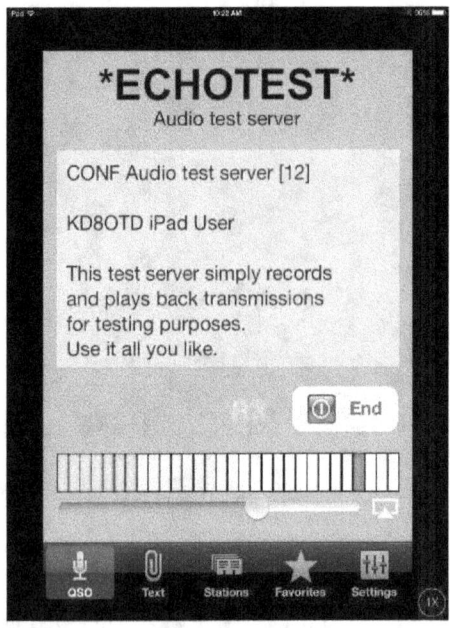

Figure 24 Echotest Activity

If it didn't work, then it's time for some troubleshooting. Go back and make sure your microphone is enabled and that you have your audio volume turned up loud enough to hear. If you accidentally told iOS not to use the microphone, you can change that: Open up the iPhone/iPad Settings App and go to Settings> Privacy > Microphone > Echolink to enable the microphone access. Android systems will have something similar.

Let's take a look at the five main screens available on

Echolink. Each of these is accessed from the buttons on the bottom of any screen:

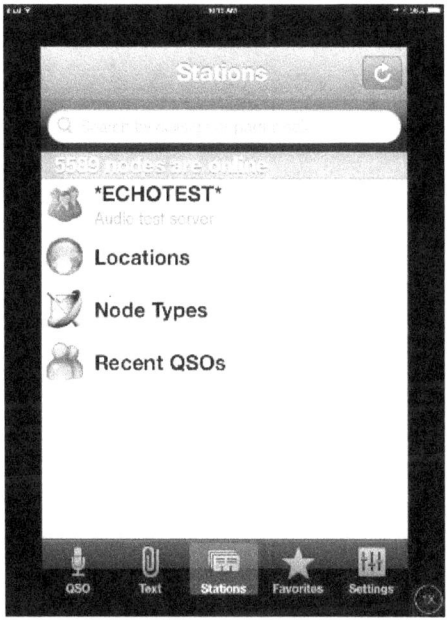

Figure 25 Five Main Control Buttons (at bottom)

QSO

If you click on the "QSO" screen right now, you'll see a message that says "Not Connected." When you are actually connected and talking to someone, this is the screen you'll be looking at. The line at the bottom of the screen is a volume slider. You can make the other party louder or quieter with this. Just above that is the level meter. When someone talks, you can see how powerful their audio is. Above that is the main window where you will see information about the person you're talking to.

Text

There's not much here when you aren't connected to anyone, but when you are connected, this serves a couple of purposes.

When you are linked to someone one-on-one, you can type messages and send them little notes while talking to them. This is a lot like texting someone on your phone, but you do it during your connection.

If you are in a conference, or group chat, sometimes people will send messages to each other while someone else is talking. Sometimes this is the group moderator, sometimes it's the listeners commenting on what's being said. It's a very flexible and convenient way for communicating, especially when someone long-winded is hogging the mike!

Stations

This is where you find someone to talk or listen to. You can search for a specific user by call sign here if you know who you want to connect with, just enter their call sign in the search box at the top of the screen and select them from the list of call signs that appears.

If you don't have a specific person in mind, you can either narrow things down by **locations** or **node types**.

Under **locations**, you can drill down by continent. Choose Africa, Asia, Europe, North America, Oceania, or South America. Then choose a specific country or region from the next list that appears. After you've chosen a country, then you can pick a specific user, link, or repeater within that country.

The other option is to choose **node types**. This brings up another menu where you can choose from Conferences, Links, Repeaters, or Users. These four options are explained in Chapter 5, "Using Echolink."

Favorites

Again, this is blank when you first start out, but once you've connected to a few conferences or friends, you'll find that there are links you return to over and over. You can save these by adding them to your "Favorites List." Think of it as your Contacts or Address Book for Echolink. It's not some-thing you absolutely have to use, but it's nice if you end up using Echolink a lot.

Settings

Assuming you have Echotest working, it's time to finish your set up. Click on "Settings" in the bottom right-hand corner of the screen. You'll get a screen that looks like this:

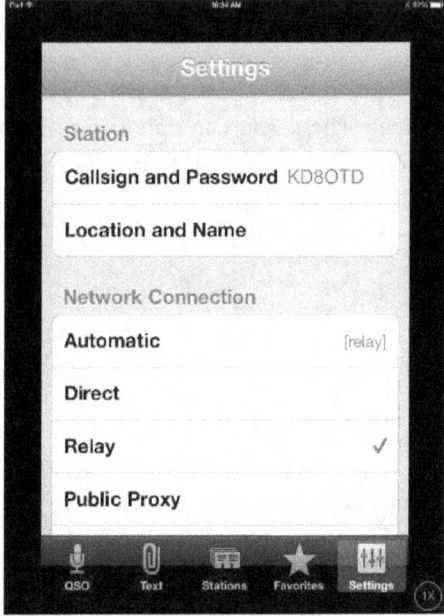

Figure 26 Echolink Settings (iPad)

Your callsign should already be in here, but you should go in and fill in your name and location. The "Network Connection" settings should not need any adjustment if you were able to get the Echotest working properly. If your network connection requires some special settings, this is where you enter it. The final two options, "End-RX Sound" and "Text Sound" are optional audible warnings that you hear as you work with Echolink. Leave them alone for now, but if the beeping sounds start annoying you later, just remember that you can turn them off.

And that's pretty much the whole Echolink system on the iPad, iPhone, and Android system. You generally start out under the **Stations** or **Favorites** menus, until you find someone to talk or listen to, then switch over to the **QSO** and maybe **Text** screens as you talk to the other party.

USING ECHOLINK

Exploring Echolink
For the following chapter, I'll be using screenshots from the PC version of Echolink, but if you are running Echolink on an Apple or Android device, all the same features are present. On the iOS/Android version, just click on the main menu at the bottom of the screen (see figure 25), and everything mentioned here should be easy to find.

From the main screen, there are two tabs, "Index View" and "Explorer View." Click on "Explorer View"

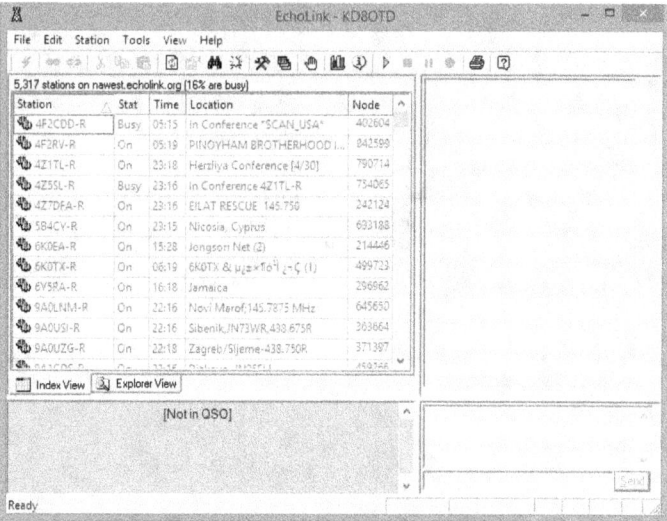

Figure 27 Echolink Main Program Screen

Which gives you a screen that looks like this:

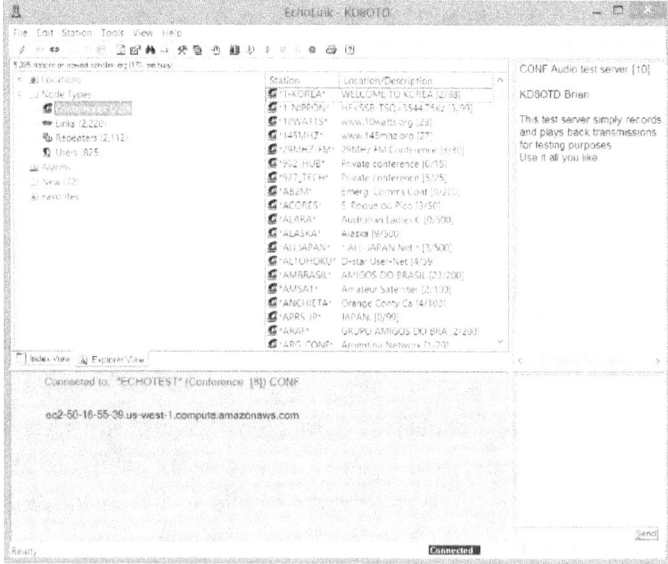

Figure 28 Explorer View, Conferences Listing

Locations

The **Explorer View** (Called **Stations** view on mobile devices) generally looks like figure 28. In the top-left window, there are a number of things you can "Explore." Locations show all the places around the world that are currently on Echolink. Want to talk to someone in Belgium or Zimbabwe right now? This is where you can find them.

Click on the "Locations" list and browse through the various countries, states, etc. that are listed there. The number in parenthesis after each listing is how many nodes are in that location. The number of nodes in the USA is usually in the thousands. Most European countries have numbers in the hundreds, but African countries often have far less.

Conferences

The second option on the Explorer list is "Node Types." There are four node types: Conferences, Links, Repeaters, and Users. If you aren't already there, click on "Conferences" to see the screen in figure 28.

Conferences are much like group chat rooms, with several people taking turns talking. In the screenshot in Figure 28, there are 228 conferences currently active, and you can scroll through them in the Explorer. The conferences usually have cryptic-sounding names, but they also have a location/description in the second column. Many of these are not in English, and are sometimes just as cryptic as the station name.

Each conference is followed by two numbers. For example, the conference named *EA3SPAIN* is described as "Echolink Espana [37/1000]" This means there are 37 people in the "chat room" and there is an upper limit of 1000 people allowed in the room. Sometimes, as in the case of *ERC-ECS*[0/500], there is no one in the conference right now.

Links

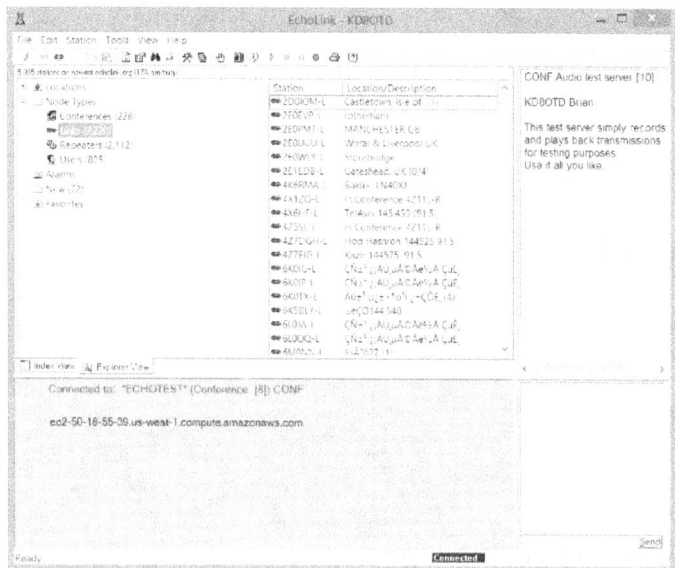

Figure 29 Links View

Links are Echolink stations out on the Internet some-where that are connected to radios. If you connect to a link and start transmitting, whatever you say will be broadcast through the radio on the other end.

In theory at least, if you connect to 2EOPMT-L and hit your spacebar, then 2EOPMT's radio will broadcast your words to other people in Manchester, Great Britain. If someone in Manchester wants to respond to you, they can use their radio to talk to 2EOPMT, and their words will come back to you through the Internet. It's like a super long-distance repeater!

In reality, as we'll see in the next chapter "Sysop Mode," many, if not most, of these links are *very* low power (like one-watt or less) and may be unattended. These links are mostly used to connect someone else's handheld radio radio to their computers remotely. If you want to talk to people in far-off places, you would probably be better served using a repeater.

Repeaters

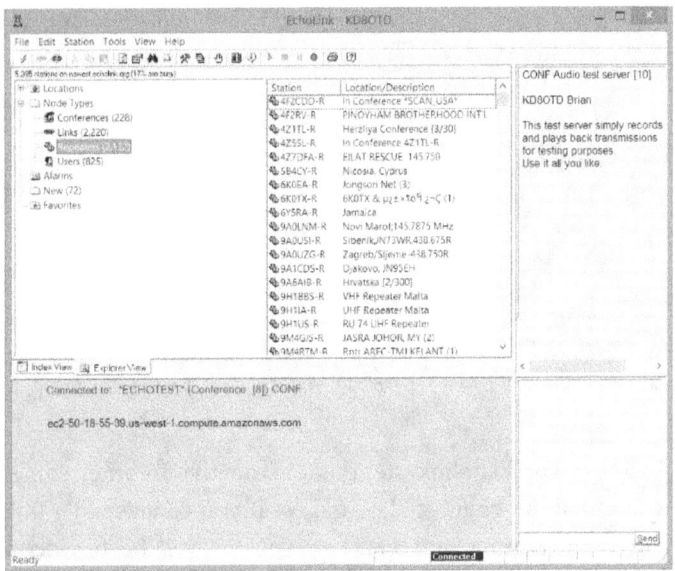

Figure 30 Repeaters

Repeaters are similar to Links, except that instead of someone's personal radio being interfaced to the Internet, this time it's a (relatively) wide-range physical repeater system just like the repeaters commonly used by UHF/VHF hams already. There's not a lot of conceptual differences between Repeaters and Links on Echolink other than the fact that repeaters are generally busier and have greater range and transmission power.

Users

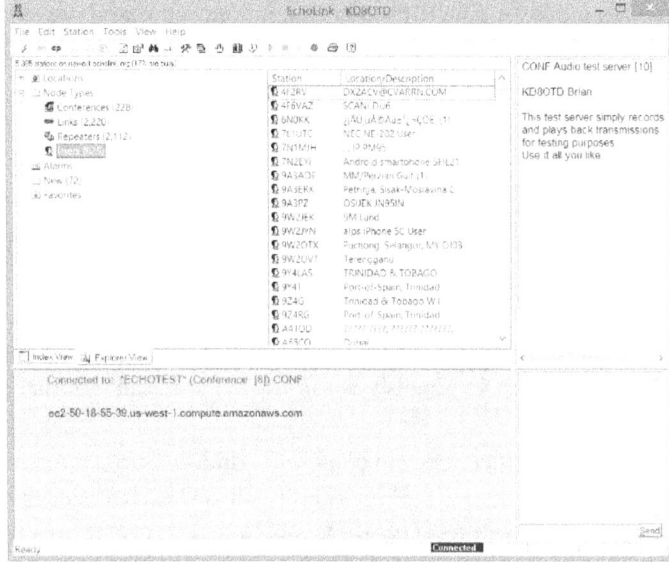

Figure 31 Users

Users are individuals who are connected to Echolink through the computer or mobile device. You can select a user from the list and talk to them directly (and privately), much like using Skype or making a phone call.

So Where do I Start?

Much like amateur radio itself, part of the fun of Echolink is that there are so many things you can do. One of the most daunting things is knowing how to get started. I would recommend digging around in the section where the Conferences are listed.

It might take some digging to find one that is in English and has people in it, but there's usually someplace with activity. A couple of English-language conferences that I usually see with someone in them are:

DODROPIN
HANDIHAM
HI-GATE
OPENIRLP
USA

And others come and go on a regular basis. Just because a conference isn't in English doesn't mean you can't connect and only listen.

Pick out a nice friendly conference and link up. Listen in for a while, and quite often they'll call on you to speak. I especially like the DODROPIN conference. They're usually active, and the people there are great with newbies.

After you've spent some time in a conference, you might ask one or two of the participants if they'd like to talk privately. Once you've arranged that, you can connect to their call sign (via the "Users" section) and ragchew all you like.

If you have "Real Radio" (not on Echolink) friends in other parts of the world, it can be a lot of fun trying to connect with them through Echolink. Link to a repeater in their part of the world and see if you can connect with your friend. **Repeaters** and possibly **Links** are great for this kind of activity.

Don't have anyone in mind, but still want to speak with someone in a faraway place? Look up a **Repeater** by **Location** and try calling CQ. The protocols work pretty much like doing a DX CQ on the regular radio.

Echolink Etiquette Tips

One piece of Echolink etiquette you should be aware of is that whenever a person joins a conference, their name goes on a list of people in the conference. It's not unusual to go into a conference server just to listen, and have someone there call on you to speak. If you want to join in, then feel free, but it's generally not a requirement. It might be nice if you answered back "Just listening tonight" or something to that effect. You can also respond in the chat window ("text" mode") if you really don't want to speak or you are having microphone problems.

A second Echolink etiquette item you should note is that when you link to a Repeater or Link, the system will announce your connection audibly to whoever is listening. Like if you've ever been talking on a regular ham radio repeater and heard something like "KD8OTD Link Connected." This is normal, and people have come to expect it. That being said, if someone connects to a repeater, they too may be expected to answer a call.

These "Link Announcements" can get very annoying to hear them over and over, so it's generally good etiquette not to link to a repeater and then disconnect and link to another repeatedly in a short time period. This in-and-out "Station Hopping" is considered bad form by ham radio listeners. Stick around a minute or two and make sure there's no activity before you abandon the link. On a related note, don't connect to a repeater and start calling CQ or whatever immediately; there could be a conversation already underway that you're interrupting. Link. Wait. Listen. Transmit.

Enhancements to Echolink

Generally speaking, the basic Echolink software is all you will ever need to connect to Echolink. If you decide to connect your radio to the Echolink system, you will do this through the same basic software. Still, some individuals have extended the interface to the software to add new features and different "skins" to the basic Echolink software.

Two of these are Echolink 100 and Artie-HF (http://qrz.com/db/N8AD). Both are well-made enhancements to Echolink. Unfortunately, at the time of this writing, neither has been updated to work with modern versions of Windows. If you are running an older version of Windows (they are untested with Windows 8 or higher), these are two systems worth taking a look at. Hopefully, their creator will eventually update them so they will work on today's systems.

Figure 32 Echolink 100

ADDING A RADIO: SYSOP MODE

For many people, what we've already covered is enough to last a lifetime. You can use your computer or mobile device to talk to hams all over the world, for free other than the cost of Internet access. What could be cooler than that?

Adding an actual radio!

Back in the first chapter, I mentioned that there was some controversy over whether or not Echolink is "real radio." If you actually have a radio hooked into the system, no one's going to be able to argue that point with you.

There are two basic ways you can use a radio combined with your Echolink software: Repeater mode and Link mode.

With Repeater mode, you set up the Echolink software so that a local repeater can communicate through Echolink; someone within a few miles of the repeater can call the repeater and use it to link to Echolink stations, repeaters, or conferences around the world. Unless you are working with a club or have special permission, most people would never have use of the Repeater mode.

The other method, Link mode, simply connects a small radio such as a handheld unit to Echolink for your own short-range use. The following chapter goes into detail about setting up a basic Linked-mode system, but there's not much difference between the setup of the Link or Repeater modes.

You'll need some kind of hardware to interface your radio to your PC. This may be as simple as a USB cable for a very modern radio, or (more likely) an extra interface box, such as a SignaLink interface or TNC. I strongly recommend the SignaLink USB if they make one specifically for your radio. You simply plug one or two cables into your radio and a USB cable into your PC, and you are good to go. They're very simple in concept. Whether or not you decide to go with the SignaLink or some other kind of linking device, you will need *something* to connect your PC and your radio. Generally, if you have an interface (TNC or sound-card style) that will allow you to operate digital modes such as PSK31 or JT65, then you probably already have everything needed.

Figure 33: SignaLink USB

Unfortunately, there are too many types of interfaces and ways of hooking up your radio to explain them all here, and new methods are introduced regularly. If you are handy with electronics, you might even try creating your own computer-to-radio interface. Related to the vast array of connection options is the even greater array of system settings that you will need to set depending on that interface.

Because of all the hardware options, the setup to get Sysop mode working in Echolink is too complex to cover every possibility, but we can look at the one that works for me. It's simple and also one of the least expensive variations. It's also generic enough to be easily modified to suit most other radio and interface combinations. This is a big part of the fun of the hardware side of Echolink—making it work!

Echolink does have an outstanding page concerning Sysop mode, which explains all the options and settings, here:

http://www.echolink.org/Help/sysop.htm

What kind of radio do I need?

Generally speaking, any radio that works with your computer interface would be fine. Most Echolink users use a UHF or VHF handheld radio, but there's no reason you couldn't use HF as well, at least in theory. In practice, you should use a radio frequency that *all* hams have legal access to, which means, in the USA for example, going with a frequency that is legal for users with a Technician-level license. This limits the frequencies somewhat. Again, most hams have access to UHF/VHF in the 144, 220, or 440 MHz bands (check restrictions for your country, of course). As far as the transmitting power of the radios, it depends on what you want to do.

If you are going to set up a link between the radio and your computer for your own use, then a low power handheld radio would be fine. If you want to cover a wider area and be used by a large group of people, the more power the better. Again, you have many choices.

Interfaces

TNC interfaces allow many types of digital use, from JT65 to PSK31 to Winmail and Packet BBS use, so they are well worth the investment if you are interested in computer interfacing or digital modes. That being said, simple sound-card interfaces such as the SignaLink allow a great deal of experimentation if you are willing to work a little harder.

For the remainder of this chapter, I will discuss the most

inexpensive, reliable option which I am aware of that uses off-the-shelf equipment.

Radio: I will use the Baofeng UV5R radio, available pretty much everywhere as of this writing. Just about any ham radio shop, or even online retailers like Amazon, have these anywhere from $35 to $50, depending on the extras that come with it. These also come with a charging cradle which will allow you to leave your radio on 24x7 if you so desire— if you plan to allow others to use your Echolink Linked station, this is something to consider.

At this point, let's assume you've already installed the Echolink PC software and gotten it working as described earlier in the book. For the next step, you will plug in the USB cable from the SignaLink (or whatever interface) into the computer, and the other cable plugs into the appropriate socket of your radio. That's pretty much the extent of the wiring portion. Easy!

Next comes the configuration, which may not be so simple. If you are running a modern version of Windows, then the SignaLink should be automatically detected as a new sound card device, called **USB AUDIO CODEC**, by your computer. If this doesn't happen, you may need to update Windows or install a sound card driver, but it's unusual that there is a problem in this step.

The first thing to do is go into Windows Control Panel and set your default sound card for Playback and also for Recording to the device "USB Audio CODEC." This tells Windows to send every output sound to the SignaLink, and also to have it listen for incoming sounds from there as well. Assuming your system recognized the SignaLink, this is all that should be necessary to configure in the Windows oper-ating system itself; everything else will be configured inside the Echolink program.

Echolink Setup

Start the Echolink software and access the "Tools -> Setup" menu option for the System Setup dialog box. The first screen is My Station, and you will need to change the MODE option to "Sysop" away from the single-user option. You will also need to change your callsign so that it ends with either – R if you are connecting to a repeater or –L if you are just linking to your own radio (which is what we are doing now now). I changed mine to KD8OTD-L:

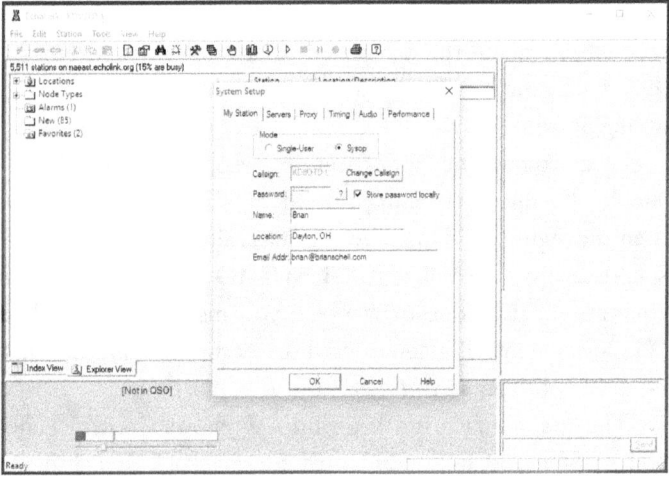

Figure 34 Changing Callsign to -L

Leave the settings in the Servers, Proxy, and Timing tabs alone. You shouldn't need to change these settings unless you have a good reason to. In the Audio tab, you will need to change both the Input Device and Output Device to the new USB Audio CODEC device to match the settings from Windows.

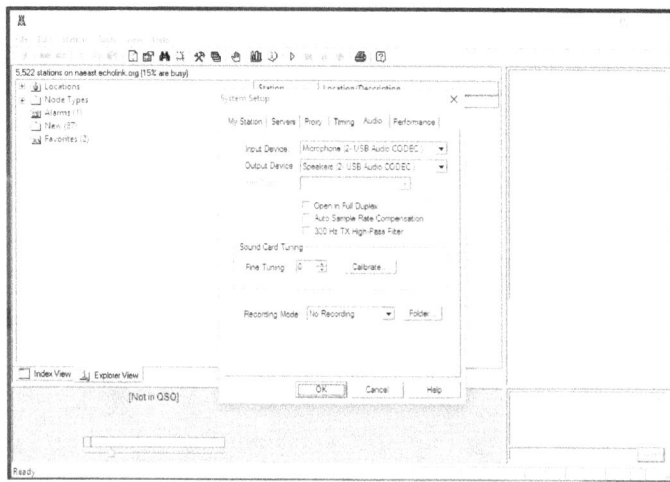

Figure 35 Settings for the Sound Card

This tells Echolink to use the SignaLink for communication. You can also experiment with the Recording Mode options at the bottom of the dialog if you wish.

Under the Performance tab, I set my Network Buffering and PC Buffering to the minimums.

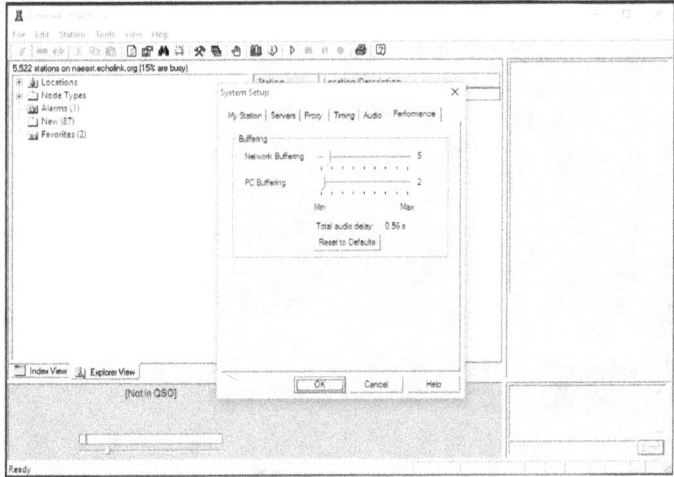

Figure 36 Timing and Buffering

My computer and Internet connection are pretty fast, so the less buffering and delay that my system gives, the better. If your system is slow or cuts off some of the words of your transmission/reception, you may want to experiment with these two settings.

Echolink Preferences

Next, go to the "Tools->Preferences" dialog box. You don't *need* to change anything under the List tab, but you may want to customize things to your liking here.

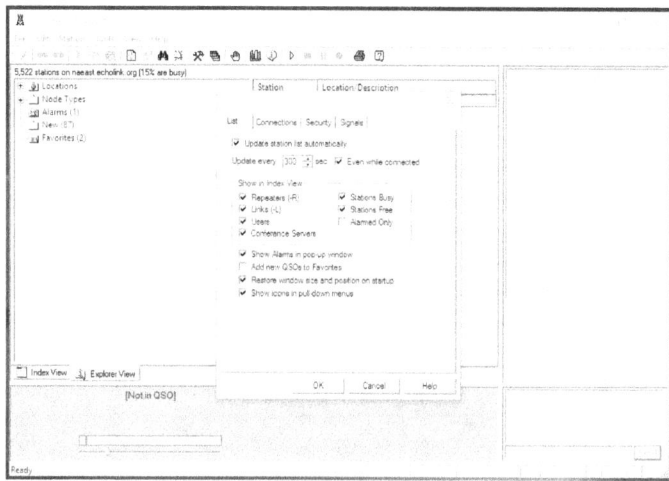

Figure 37 The List Tab

There are several changes that need to be made under the Connections Tab. Click on the button for "PTT Control." Depending on your TNC or interface, you may need to set this to whatever your COM port requirements are. If you are following along with me with a SignaLink, the SignaLink doesn't use a COM port, it simply works with sounds. You will uncheck the "Serial Port CTS" option if it is checked already.

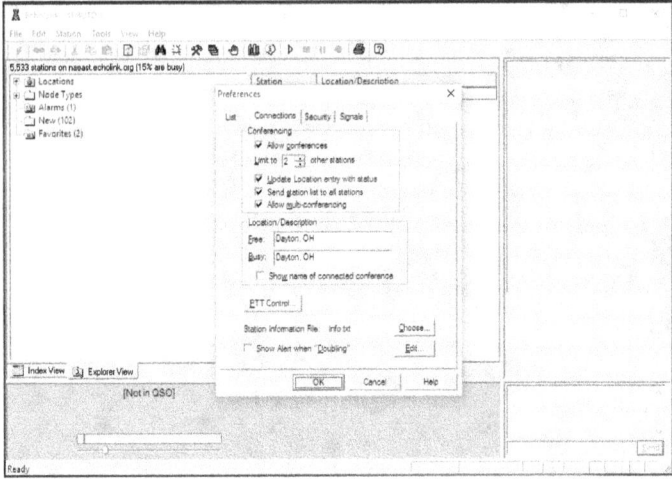

Figure 38 Connection Options

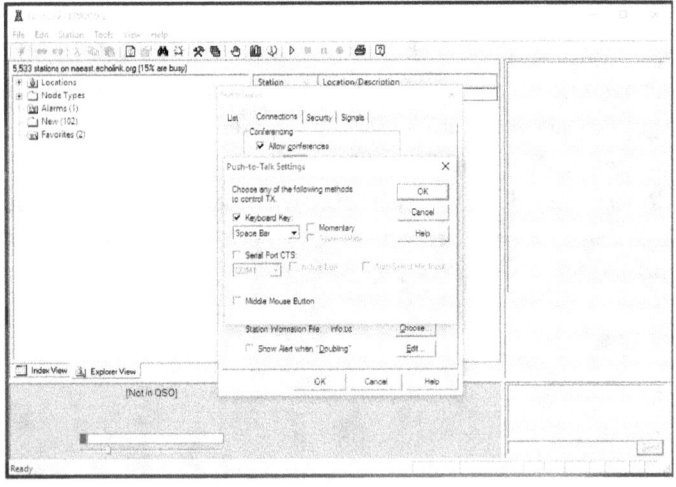

Figure 39 Push-To-Talk Options

It's probably not necessary to make any changes to the Security or Signals tabs. If you want to refuse connections to or from certain countries, you can easily set that up here.

Echolink Sysop Settings

The third major dialog we have to fix is under "Tools->Sysop Settings."

Under the RX Ctrl dialog, I have mine set for the SignaLink with Carrier Detect = VOX and the following settings:

> *VOX Delay = 1500*
> *Anti-Thump = 1500*
> *ClrFreq Delay = 1000*

Try these numbers to start, and if you find that your radio cuts off portions of transmissions or if it sits there doing nothing and wasting seconds, then you may want to experiment with fine-tuning these delays later.

These settings work well with the UV5R and SignaLink. If you are not using the SignaLink, then you may need to turn on the COM port option or adjust your settings in other ways.

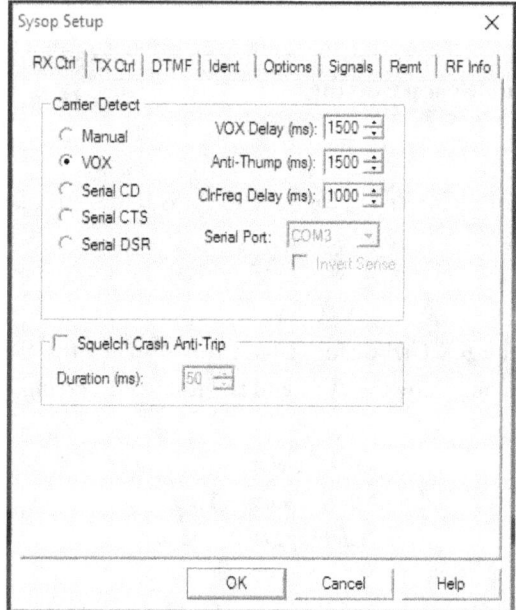

Figure 40 RX Control Options

On the TX Ctrl dialog, the SignaLink uses External VOX for PTT Activation. Other interfaces may once again need to use a COM port.

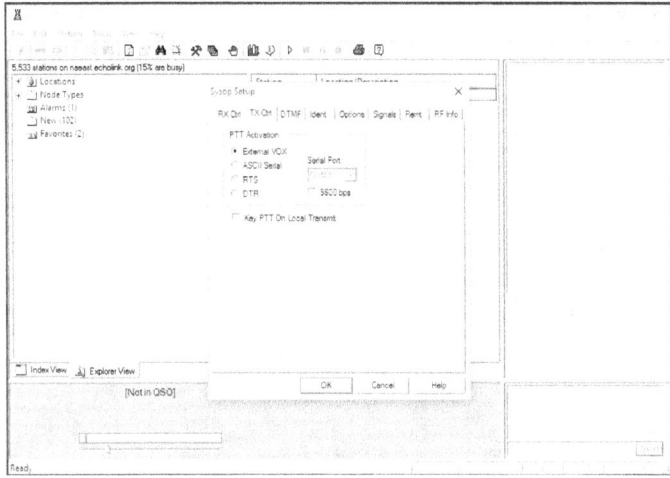

Figure 41 TX Control Options

You don't need to change anything immediately on the DTMF dialog, although once again, these can be customized later if you prefer.

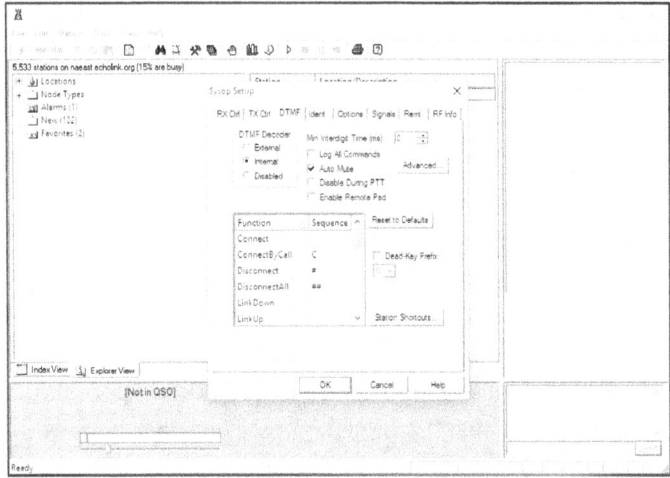

Figure 42 DTMF Options

Under the Ident tab, you'll need to change your call sign to end with either −R or −L again. You can configure Echolink to identify your station in either Morse code or by spoken voice (recorded in an external file) here. You can set your station identification delays and settings on this screen as well. I have mine set to play the Morse code every 6 minutes. This is not strictly required as long as you identify yourself according to the rules of amateur radio, but it doesn't hurt to play it safe.

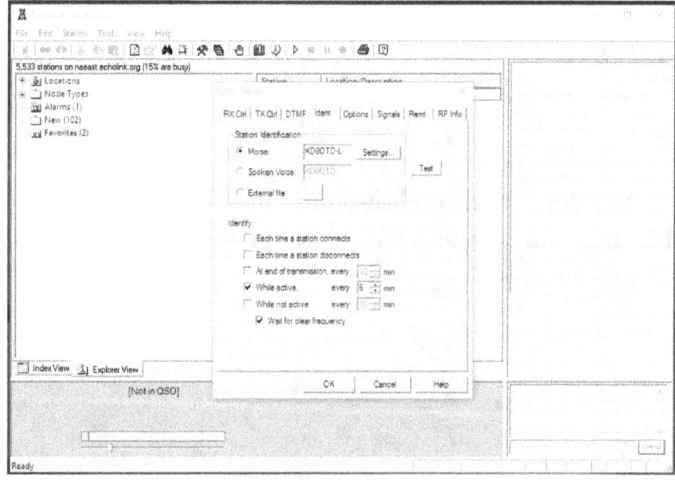

Figure 43 Ident Options

For my setup, I didn't see much need to change anything under the Options, Signals, or Remt tabs, but your mileage may vary. Under the RF Info, you can put your latitude and longitude in, as well as the frequency that your radio is tuned to, which we'll discuss a little later. As you can see from the picture below, I have chosen 147.555. The other options on this screen can be set however you desire.

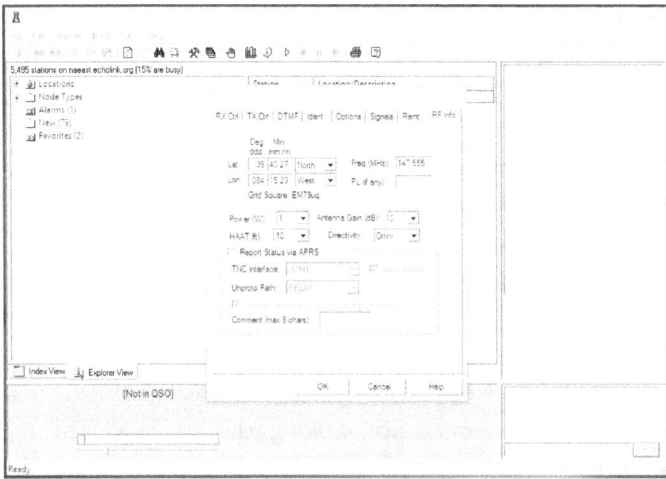

Figure 44 RF Info Settings

In theory, that's all you need on the software side of the project. Now onto the radio itself. The radio only needs a few settings changed. First, I changed the VOX setting almost to the most sensitive, level ten. The Baofeng UV-5R doesn't have an auto-power-off feature, so if you leave it in the cradle, it will run forever. If you are using some other radio, you will need to change the auto-off timer to "Never."

And now you need to set a frequency. As I mentioned earlier, most of the time Echolink users will use VHF or UHF handhelds to connect, so that usually means choosing a SIMPLEX frequency in one of the VHF or UHF bands. See "Appendix B: Common VHF/UHF FM Simplex Frequencies" to help you select a frequency. Don't choose the National Calling frequency for your band, use one of the lesser-used frequencies. I chose 147.555 for mine, since that frequency is well down on the list and not often used in my area.

Additional Information about Repeater Mode:

The previous information about choosing a frequency is assuming you are setting up an Echolink **Link** (Just your own radio and a computer). If you are setting up Echolink for use with a club repeater, then you'll need to set the radio to transmit and receive on whatever frequency pair and with whatever offset or codes that repeater requires, just as if you were going to use the radio to talk through the repeater. If you have physical, local access to the actual repeater hardware, you could also directly wire the repeater into your computer for permanently linking the repeater into the Echolink system. For that, check the repeater manual for details on how to interface the repeater to external devices.

Testing the Echolink Link

Make sure the Echolink software is running and the radio is turned on. Click on "Tools->Sysop Settings->Ident" and then click on the Test button. If everything is correct, your radio should activate and transmit your station ID either in Morse or using the recording you set up over the airwaves.

If it works, congratulate yourself, and if it doesn't, then it's time to backtrack and narrow down which setting is causing the problem. At this stage, if the radio doesn't transmit, then it's most likely a problem with the settings relating to PTT (Push to Talk) not being correct under "Tools->Sysop Setup->TX Ctrl" and/or "Tools->Preferences->Connections." Double-check your settings and experiment with other settings if something doesn't work.

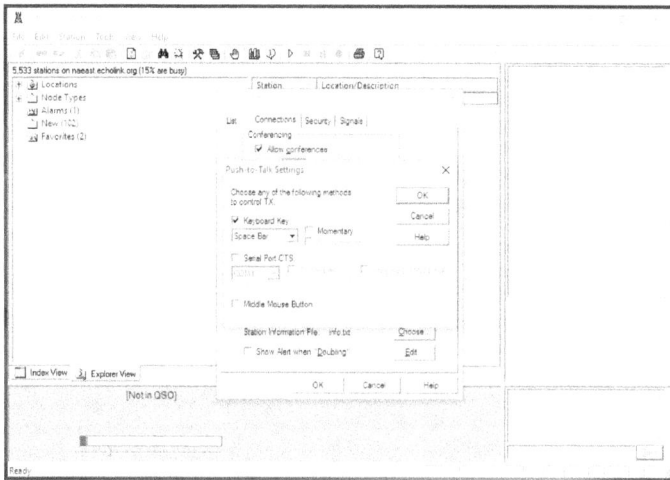

Figure 45 PTT Settings

Once you have verified that your computer can access your radio and transmit through it, it's time to test it from the opposite direction. Using a second radio, tune it to the needed frequency (147.555 in my case again), and say your call sign. The radio connected to your computer should activate when it picks up your transmission. If it doesn't, check that the frequencies on the two radios are set similarly. You don't need to set up CTCSS or tones or anything (this is simplex after all), just a plain frequency is fine, just make sure they match exactly.

On the microphone or handheld radio that you are using for testing, press the * key. You *should* hear the Echolink system broadcast a status message back to you. Mine simply reads back my callsign, "KD8OTD." If the connected radio seems to be working properly, but the Echolink software doesn't seem to "hear" anything, check your settings under "Tools->RX Ctrl."

Assuming you got the status message, it's time for the big test. On your microphone keypad or HT radio, key in the number "9999" (four nines). You *should* now be connected to the Echolink Test Server. You should hear the standard message about recording a message which will be played back to you. Transmit your message and listen for the response. If the Echolink system then replays your message back to you, you are done, and now have a working Echolink Linked system!

Now that you have a Link station, what do you do with it? That's the subject of the next chapter.

USING ECHOLINK WITHOUT A COMPUTER

R adio-only Operation

In this chapter, I am assuming you are going to connect to either your own or someone else's Linked Echolink system with your radio. If you just set up your own link in the previous chapter, now we'll discuss how to control it with only your handheld HT radio. You can also use and control clubs' and other people's linked systems.

Linking with DTMF Tones

If all you have in your hands is a handheld HT radio, you don't have a huge array of controls to work with, so the powers-that-be have come up with a system that allows you to input commands with only the little numeric keypad on the radio. See **Appendix C** for a table listing all the command codes.

In the previous chapter, we pressed the * key to receive an ID message from the Echolink system. Try that again to make sure everything is still set up correctly. Just tune your radio to whatever frequency you need and press the * key on the radio's keypad (or on the microphone, or wherever you have a keypad). If you set up your own link, you should hear the name of your link (your callsign) read back to you.

There are many commands beyond simply getting a status message.

Looking back at the main Echolink screen for a minute, you can see that each station listed has a "Node" number. This is how Echolink identifies stations. To connect with one of these other stations, simply type in the node number on your handset keypad. Once connected, you can talk just as if the other person was within range of your radio.

When you are finished, you simply enter # on the keypad to disconnect. There are many other options that you can control through the keyboard, such as entering a station by its callsign, choosing a listen-only mode, as well as the ability to look up node numbers and callsigns. See **Appendix C** for the whole list of commands.

APPENDIX A: REGISTRATION

So now you have your copy of Echolink installed and are ready to play, right?. Not so fast!

Before your can get on the air, or even on the Internet, with Echolink you need to validate that you are legal to use the system. Remember, that even if you are talking into your computer or iPad, depending on who you are talking to, you might still be talking over the airwaves via radio; sometimes you can't even tell from your end. Since there is always the possibility of your voice hitting the airwaves, the FCC and other Radio "Powers That Be" require that you have a valid Amateur Radio license. Now it's time prove it.

The first step is to download, install, and run the Echolink software, and set it up as we have already done in Chapter 2. When you entered your call sign in the program, it was automatically registered with the Echolink servers.

After you've got the program run and set up, go to the website: http://www.echolink.org/validation/ and follow the instructions from there.

Probably the fastest way to get validated is to scan and submit the documentation they need. The required documentation varies depending on what country you are in. In the USA, they need a photocopy or scan of your ham radio license. This is the actual paper copy of your license. If you aren't in the USA, other documents may be acceptable.

See http://www.echolink.org/validation_docs.htm for the full list of acceptable documents.

Don't have a scanner? Take a clear photograph of the license with your digital camera or cell phone and submit that. This must be a copy of your actual FCC license, not a download off a website or anything else.

Why is all this necessary? There are many logbook sites like http://qrz.com that the Echolink people could use to verify that you are a licensed ham operator. The point is that this requirement is to verify that you are who you claim to be. If someone gets on Echolink and claims to be me, KD8OTD, there's no doubt that it's a valid call sign, but is it really me submitting it? They have no way of knowing without some kind of proof. It's inconvenient, but it only has to be done once.

I know in my case, the worst part was just finding that little slip of paper called a license. I know it's around here *somewhere...*

APPENDIX B: COMMON VHF/UHF FM SIMPLEX FREQUENCIES

2-Meter Band	1.25-Meter Band
146.52*	223.42
146.535	223.44
146.55	223.46
146.565	223.48
146.58	**223.50***
146.595	223.52
147.42	
147.435	**70-cm Band**
147.45	
147.465	**446.0***
147.48	
147.495	
147.51	**33-cm Band**
147.525	
147.54	**906.5***
147.555	
147.57	**23-cm Band**
147.585	

23-cm Band

1294.000
1294.025
1294.500*
Every 25 kHz to 1295

* National simplex calling frequencies

Common VHF/UHF FM Simplex Frequencies

APPENDIX C: DTMF FUNCTIONS

Command	Description	Default
Connect	Connects to a station on the Internet, based on its node number.	num
Connect by Call	Connects to a station on the Internet, based on its callsign.	C+call+#
Random Node	Selects an available node (of any type) at random and tries to connect to it.	00
Random Link	Selects an available link or repeater (-L or -R) at random and tries to connect to it.	01
Random Conf	Selects a conference server at random and tries to connect to it.	02
Random User	Selects an available single-user station at random and tries to connect to it.	03
RandomFavNode	Selects an available node (of any type) at random from the Favorites List and tries to connect to it.	001
RandomFavLink	Selects an available link or repeater (-L or -R) at random from the Favorites List and tries to connect to it.	011
RandomFavConf	Selects a conference server at random from the Favorites List and tries to connect to it.	021
RandomFavUser	Selects an available single-user station at random and tries to connect to it.	031
Disconnect	Disconnects the station that is currently connected. If more than one station is	#

DTMF Functions (More on next page)

	connected, disconnects only the most-recently-connected station.	
Disconnect All	Disconnects all stations.	##
Reconnect	Re-connects to the station that most recently disconnected.	09
Status	Announces the callsign of each station currently connected.	08
Link Down	Disables EchoLink (no connections can be established).	(none)
Link Up	Enables EchoLink.	(none)
Play Info	Plays a brief ID message.	*
Query by Call	Looks up a station by its callsign and reads back its node number and status.	07+call+#
Query by Node	Looks up a station by its node number and reads back its callsign and status.	06+num
Profile Select	Switches to a different stored set of configuration settings (0 through 9).	B#+num
Listen-Only On	Inhibits transmission from RF to the Internet.	0511
Listen-Only Off	Restores normal transmission from RF to the Internet.	0510

DTMF Functions (Continued)

These are included here for convenience, but have been taken from http://www.echolink.org/Help/dtmf_functions.htm

Connect

The default for the Connect command is to simply enter the 4- 5-, or 6-digit node number to which you wish to connect. To prevent interference with other DTMF functions, however, you may wish to configure a special prefix, such as A or 99.

Link Up and Link Down

No defaults are provided for these functions. To enable

these functions, enter a DTMF sequence for each one, using the DTMF tab of the Sysop Settings page.

Profile Select

Profiles are numbered from 0 to one less than the number of profiles shown under File->Profiles. Profile 0 is always MAIN.

Station Shortcuts

Custom DTMF commands can be created to connect to specific stations. These commands are called Station Shortcuts, and are not shown in the table above. To manage your Station Shortcuts, click the Station Shortcuts button on the DTMF tab of Sysop Settings.

Entering Node Numbers

To enter a node number (for the Connect or Query by Node commands), enter the 4-, 5-, or 6-digit node number. If the specified node is not among the stations currently logged on, EchoLink will say "NOT FOUND".

Entering Callsigns

To enter a callsign (for the Connect by Call or Query by Call commands), press two digits for each letter and number in the callsign. The first digit is the key on which the letter appears (using 1 for Q and Z), and the second digit is 1, 2, or 3, to indicate which letter is being entered. To enter a digit, press the digit followed by 0. When finished, end with the pound key (#).

For example, the letter "K" is entered as "52", the letter "Q" is entered as "11", and the digit "7" is entered as "70".

Callsigns need not be entered in full. If a partial callsign is entered, EchoLink will find the first match among the stations currently logged on. If no match is found among the stations currently logged on, EchoLink will say "NOT FOUND".

Examples

(These examples assume that the default DTMF codes are configured in Sysop mode.)

To connect to node number 9999:

Enter: **9 9 9 9**

EchoLink responds with:

"CONNECTING TO CONFERENCE E-C-H-O-T-E-S-T"

followed by

"CONNECTED"

because 9999 is the node number of conference server "*ECHOTEST*".

To get the status of K1RFD:

Enter: **0 7 5 2 1 0 7 2 3 3 3 1 #**

EchoLink responds with:

"K-1-R-F-D 1-3-6-4-4 BUSY"

because 13644 is the node number of station K1RFD, and K1RFD is currently busy.

To connect to a random link or repeater:

Enter: **o 1**

EchoLink responds with:

"CONNECTING TO K-1-O-F REPEATER"

followed by

"CONNECTED"

because K1OF-R was selected at random.

ABOUT THE AUTHOR

Brian Schell (KD8OTD) is a former College IT Instructor who has an extensive background in computers dating back to the 1980s. Currently, he writes on a wide array of topics from computers, to world religions, to ham radio, and even releases the occasional short horror tale.

He'd love to hear your stories of success and failure with working with Echolink. If there's something you would like to see in a future edition of the book, or otherwise have suggestions, please drop him a note. Contact him at:

Web: http://BrianSchell.com
Email: brian@brianschell.com

twitter.com/BrianSchell

facebook.com/Brian.Schell

instagram.com/brian_schell

pinterest.com/brianschell

ALSO BY BRIAN SCHELL

Amateur Radio

• D-Star for Beginners

• Echolink for Beginners

• DMR for Beginners Using the Tytera MD-380

• SDR for Beginners with the SDRPlay

• OpenSpot for Beginners

• Programming Amateur Radios with CHIRP

Technology

• Going Chromebook: Living in the Cloud

• Going Text: Mastering the Power of the Command Line

• Going iPad: Ditching the Desktop

• DOS Today: Running Vintage MS-DOS Games and Apps on a Modern Computer

The Five-Minute Buddhist Series

• The Five-Minute Buddhist

• The Five-Minute Buddhist Returns

• The Five-Minute Buddhist Meditates

• The Five-Minute Buddhist's Quick Start Guide to Buddhism

• Teaching and Learning in Japan: An English Teacher Abroad

Fiction with Kevin L. Knights:

• Tales to Make You Shiver